QUANTUM ANALYSIS OF PRIMARY SUCCESSION

The energy structure of a vegetation chronosere in
Hawai'i Volcanoes National Park

Buried deep under masses of lava, once a thriving native Hawaiian village, Kalapana

" … the noise of the bobbling lava is not great … .
The smell of sulphur is strong, but not unpleasant to a sinner. "

Mark Twain in
"Roughing it in the Sandwich Islands." Pp. 71-72.
Mutual Publishing, Honolulu, Hawaii 1990.

Mártának
"Én nem tudom , mi ez, de édes ez,
Egy pillantásod hogyha megkeres,
Mint napsugár ha villan a tetőn,
Holott borongón már az este jön."

Juhász Gyula

About the book
The book revisits the fundamental quantum ecological idea that the potential energy structure of a plant community is a seamless fusion of footprints specific to basic processes which operate on all scales – *phylogeny, environmental mediation, and chance.* At this time the idea is tested in quantum analysis of a vegetation chronosere in Hawai'i Volcanoes National Park.

QUANTUM ANALYSIS OF PRIMARY SUCCESSION

The energy structure of a vegetation chronosere in
Hawai'i Volcanoes National Park

László Orlóci FRSC
Western University

SCADA Publishing – London, Canada

Refer to this book:

Orlóci, L. 2013. Quantum analysis of primary succession. The energy structure of a vegetation chronosere in Hawai'i Volcanoes National Park. SCADA Publishing, Canada. Online Edition:
https://createspace.com/4452597

Look for these books:

Orlóci, L. 2013. Quantum Ecology. Energy structure and its analysis. SCADA Publishing, Canada. Online Edition: https://createspace.com/4406077

Orlóci, L. 2013. On the Energy Structure of Natural vegetation. In search for community governance rules. SCADA Publishing, Canada. Enlarged Online Edition:
https://createspace.com/4153484

Orlóci, L. 2012. Self-organisation and Mediated Transience in Plant Communities. SCADA Publishing, Canada. Enlarged Online Edition: https://createspace.com/3585127

Orlóci, L. 2012. Statistical Ecology. The quantitative exploration of nature to reveal the unexpected. SCADA Publishing, Canada. Online Edition: https://createspace.com/3476529

Orlóci. L. 2011. Problem flexible computing in statistical ecology. SCADA Publishing, Canada. Online Edition: https://createspace.com/3574792

Orlóci, L. 2012. Statistical multiscaling in dynamic ecology. Probing the long-term vegetation process for patterns of parameter oscillations. SCADA Publishing, Canada. Online Edition: https://createspace.com/3830594

ISBN-13:978-1492788997
ISBN-10:1492788996

V – 2014-01-01

Look for information:
https://sites.google.com/site/statisticalecology/

QUANTUM ANALYSIS OF PRIMARY SUCCESSION
All rights reserved © 2014 by L. Orlóci & M. Orlóci

Contents

Contents 5
Preamble 6
INTRODUCTION 7
QUANTUM ANALYSIS 12
    Potential energy 12
    Further on energy-based entropy 14
    The zero data element dilemma 17
ENERGY STRUCTURE 18
    The species level 18
    Energy hierarchy 20
    Energy trends 23
    Homogeneity tests 26
DISCUSSION 31
References 36
Index 38
Appendices 41
Bibliographic notes 55

Preamble

The book revisits the fundamental quantum ecological idea that the potential energy structure of a plant community is a seamless fusion of footprints specific to basic processes which operate on all scales – *phylogeny, environmental mediation, and chance.* At this time the idea is tested in quantum analysis of a vegetation chronosere in Hawai'i Volcanoes National Park.

How is the test constructed? What are the outcomes? What do the results tell about primary succession not already known from other sources? Stated in the briefest of terms the test requires temporal species performance data. This comes from the survey of a proxy succession sere, also known as a chronosere. The test criterion is Max Planck's energy-based entropy function. This is applied in a multiscale, probabilistic manner. The results are statistical generalizations concerning the energy structure of primary succession. The book gives details and explains the results at length.

7 | Quantum analysis of succession

INTRODUCTION

The Preamble identified phylogeny, environmental mediation, and chance as the three basic processes which define the energy structure of primary succession. It should be added right at this point that in succession research we are interested in *phylogeny* because it is Nature's mechanism to create species richness and to endow species with functionality. As such, phylogeny has its energy footprint in the succession process. C*urrent environmentally mediated transience* enters considerations as Nature's mechanism by which species functional types are selected in community assembly. This too have its energy footprint in the succession process. The role of *chance* is too an object of interest by virtue of its unavoidable presence in all things happening in Nature. As such, conditions attributed to chance effects serve as threshold to which all results can be referred when significance is tested.

The Book's central theses can be stated in the form of two propositions about the energy structure of primary succession:

Proposition A. The vegetation's potential energy structure is a seamless fusion of energy footprints specific to the basic processes.

Proposition B. The vegetation's potential energy state rises in the course of primary succession.

The principle aim of quantum analysis is the isolation of the specific energy footprints and the test of hypotheses about the footprints. These require, at best, long-term observations in situ. These are rarely available. The next best evidence is in the survey records of a *spatial chronosere* which we use as the proxy succession sere. The site of a chronosere is usually the vegetation catena, oriented in such a way that it traverses vegetation communities in spatial order deemed to represent a true succession sere in time. Other sources of evidence are synthetic chronoseres based on free standing quadrats strategically selected in the manner of *space for time substitution*, as Wildi and Schüts (2000) described the sampling maneuver.

At this time the data we revisit are old phytosociological records, containing the estimated percent ground cover of species, which Márta Mihály and I collected in our 1997 survey of a catena in the Devastation Trail area of Hawai'i Volcanoes National Park (see Figure 2). The catena site (NE oriented straight line on the Google map in Figure 1) takes up part of the concave lower slope of Kilauea Iki's cinder cone. Appendix 1 contains a partial data set including records for 51 species in 23 contiguous sample plots (30m x 30m each). Three characteristic sections of the catena are highlighted by photographs in Figure 2.

Before we discuss further details, which are directly relevant to our aims, we refer the reader to a recent monograph by Mueller-Dombois, Jacobi, Boehmer and Price (2013) in lieu of an introduction. The monograph covers ecological conditions and plant geography on the Hawaiian Islands, focusing on the Metrosideros Rainforest Biome in which the Devastation Trail area is located. We refer to Smathers and Mueller-Dombois

9 | Quantum analysis of succession

(2007) for interesting facts concerning the vegetation dynamics of extreme environments within the neighbourhood of our study site, and to the monumental work of Mueller-Dombois and Fosberg (1998) which puts the Hawaiian Islands' ecology and natural history into the global context of the Tropical Pacific Islands. The article by Kitayama, Mueller-Dombois and Vitousek (1995) rounds out the general references with specific regard to primary succession on lava flows in the montane Metrosideros Rainforest Biome on the side of the giant volcano, Mauna Loa. The sampling design used by Kitayama and coauthors is an éclat case for space for time substitution, but not the catena type. They construct a composite proxy succession sere which pools information from sample plots located on the same lava type of different flows, whose ages range from less than a decade to thousands of years.

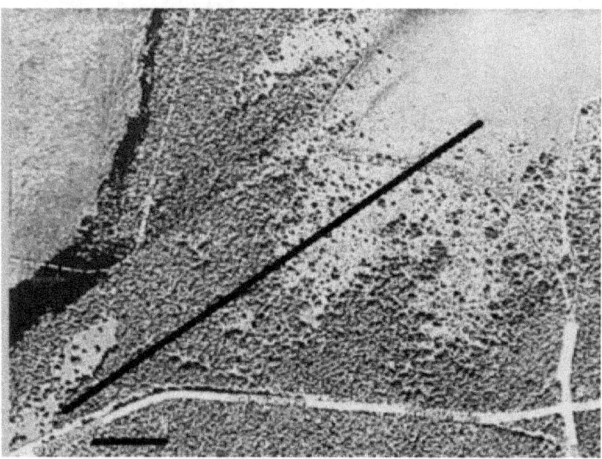

Figure 1. Google map of the Devastation Trail area on Kilauea in the Hawai'i Volcanoes National Park. The long straight line from NW to SW marks the site of the chronosere which Márta Míhály and I surveyed in 1997. Dark patches indicate vegetation. Long black patch on left is probably

shadow cast by the Halema'uma'u caldera's East wall. White patches can be reflections of dead wood. North East corner: top of Kilauea Iki cinder cone. Short black horizontal line at left bottom indicates 100 m.

11 | Quantum analysis of succession

Figure 2. Three characteristic sections of the catena on the Kilauea Iki's cinder cone. Leading species: Ecesis – Dubautia scabra, Styphelia douglasii, Nephrolepis exaltata; Graminoids and shrub – Myrica faya, Metrosideros polymorpha, Vaccinium ciliatum, Andropogon virginicum, Veronica sp.; Forest and fern – Metrosideros polymorpha, Cibotium glaucum, Hedychium gardnerianum.

Our elected objective is to test the propositions already stated. The way to do this is statistical, based on a family of multiscale techniques to which we refer as *quantum analysis*. The designation signifies that quantum theoretical principles are applied in the manner introduced by Orlóci (2013). It should pleasantly surprise the ecologist reader that the zero inundated data sets, he likely to possess, so typical for vegetation surveys, is particularly well suited for quantum analysis. The reason is in part the definition of the vegetation's potential energy state as a function of marginal totals (T) in the data table and the number of data elements (n) contributing to the marginal totals (see Appendix 1). There is nothing new in quantum analysis as far as the basics are concerned. Fundamental theory and *modus operandi* existed in the public domain since Max Planck published his paper on quantum theory in 1901.

QUANTUM ANALYSIS

The term "quantum" probably invokes memories in readers (pleasing or otherwise) from high school or college experiences in physics lectures on Max Planck and his quantum theory. Ecological quantum analysis does in fact incorporate in its core Max Planck's (1901) quantum theoretical principles. We give a brief review, from our point of view, but first for fluency in discourse we require clarification of the basic technical terms.

Potential energy

Energy can be measured in units of many types (Orlóci 2013). We remember well from introductory physics that the potential energy state of an object whose mass is M = 1 kg, lifted up to height h in meters (m) against gravitational acceleration g is given by ε = gh in mkgs^{-2} units. Let the object be one of n = 4 identical kinds. The potential energy level in the set is E = 4ε mkgs^{-2}.

If we think what we really did to get to E = 4ε mkgs^{-2}, we can establish a general principle: when we measure an object in actual fact we are counting energy units. We have now formulated the first part, the data acquisition part of quantum analysis. In further steps we need access to the value of n, the number of traits in which we counted energy units, and T, the

13 | Quantum analysis of succession

total number of energy units counted. Let us call the set of n traits a complex of n resonators.

We can adopt these conventions: X – data set; n - number of resonators; s – number of hierarchical relevés (see Orlóci 2013), one per sample plot (quadrat, catena segment). Based on these we can write E=nε, but NOT in $mkgs^{-2}$ units. If not in those units, then in what units? Since the unit in which energy is measured is completely arbitrary, we should feel free to opt for any unit we find useful for our purposes. I choose natural units (nats).

The choice of natural units takes us by way of Max Planck's (1901) reasoning to his Nobel Prize winning *energy-based entropy function*:

$$E = nH = -\ln P = \ln C = (T+n) \ln (T+n) - T \ln T - n \ln n \text{ nats}$$

in which

$$C = \frac{(T+n-1)!}{T!(n-1)!} \approx \frac{(T+n)^{T+n}}{T^T n^n}$$

by Stirling's approximation, and further,

$$P = \frac{1}{C} \text{ and } H = \frac{nH}{n} \text{ nats.}$$

The question that the contemplative reader probably wants us to answer at this point is this: what justifies using measurements as energy unit counts, and further, what justifies to use nH or H to measure the energy state of the vegetation community? We can point to two of Max Planck's quantum theoretical principles:

1. Energy is transmitted in discrete units, the quanta. Therefore energy units can be counted.

2. nH or H is an alternative parameter for the energy state of a complex of n resonators in the manner of

$$H = \left(1 + \frac{E}{\varepsilon}\right)\ln\left(1 + \frac{E}{\varepsilon}\right) - \frac{E}{\varepsilon}\ln\frac{E}{\varepsilon}$$

Complete derivation is in Planck's 1901 paper, in somewhat different symbolic term.

Regarding the terminology, consider the vegetation community of n species within one section of the catena as "a complex". The species are the "resonators". We describe the community by a hierarchical relevé (Orlóci 2013).

"Complex" and "resonator" may refer to different kinds of ecological entities. Because of the context dependence of the ecological equivalents of "complex" or "resonator", we continue using Max Planck's generic terms, "complex" and "resonator", and only shift to the ecological equivalents in the discussion where clear definition exists.

Further on energy-based entropy

Turning back to the equations in the preceding section, we see the dependence of nH and H on resonator richness n modified by resonator abundance T. C is the possible number of different arrangements of T energy units among n resonators, and P is the probability of a randomly assembled resonator complex turning out to be exactly the same as the one actually observed. Restated for a T-totalled community data set, such as the cover estimates of n species (Appendix 1), the probability of an n-species assemblage to turn out by random sorting exactly the same as the hierarchical relevé in hands within a catena segment to which the relevé applies, is exactly $P = \dfrac{1}{C}$.

A further point, when H is given, P is defined by e^{-H}. In tests of uniqueness H is considered statistically unique (the term "significant" may also be used) when it is at least 3 nats or P is not higher than 0.05. These are changeable arbitrary rules of thumb.

All fundamental terms specified and now we can state in succinct terms what do we use quantum analysis for in this paper in the first place:

We wish to measure the potential energy state of vegetation complexes. We wish to partition the total potential energy into components specific to the fundamental processes. After we isolate the specific components we can consider specific tests on the energy structure of the succession sere.

A detailed description of theory and implementations is found in Orlóci's (2013) recent essay under the title *Quantum Ecology*. For a more basic level of insightful facts the reader may turn to Stephen Hawking's *"The dreams that stuff is made of"* (2011) in which the reader can find Max Planck's 1901 paper on the theoretical underpinnings of quantum analysis.

It would not be too far from what it is intended for in Ecology, to consider Quantum analysis as a high-level energy-based diversity analysis. This fact prompts us to point out what makes quantum analysis a different kind of species, sharply isolated from the techniques we have seen developed about the physical properties represented by Rényi's (1961) generalised entropy and information functions H_α and I_α. Written for a two-dimensional contingency table,

$$H_\alpha = \frac{1}{1-\alpha} \ln \sum_{i=1}^{n} p_i^\alpha \quad \text{and} \quad I_\alpha = \frac{1}{\alpha-1} \sum_{i=1}^{n} \frac{p_i^\alpha}{q_i^{\alpha-1}}$$

(Orlóci 2006, 2012).

Rényi's functions require p and q values. For the former we have to reach back to the basic elements in the T-totalled data vector **X** and to define p as $p_i = \frac{X_i}{T}$; q values could come from theory. Note that in quantum analysis, such p or q values are not used. P is based on T and n through the quantity C.

To avoid possible confusion, we point out that the similarity of nH=lnC nats in quantum analysis and Brillouin's information equation I=$\log_2$C bits[1] superficial. Brillouin's C is defined by $\frac{N!}{f_1! \, f_2! \, \ldots \, f_s!}$ in which $N = \sum_{i=1}^{s} f_i$ and f_i is an element from which the p_i for Rényi's equation can be calculated, $p_i = \frac{f_i}{N}$. Furthermore, Brillouin's I is not a divergence measure, therefore it is not the same as Rényi's I.

---

[1] Note: $\log_2 2$ is one bit and the maximum I is $\log_2 s$ bits. Brillouin shows that when the f_i are large, say 100 each or greater, then $\frac{I/n}{\log_2 e}$ will come close in value to Shannon's (1948) entropy function. In this, *e* is the natural base (2.718281828).

17 | Quantum analysis of succession

We repeat, in Planck's nH or H the p and q parameters do not enter direct consideration. The parameterization of the energy equation relies entirely on T and n.

The zero data element dilemma

Typically for our type of vegetation data is the inundation of the contingency table by zeros. Out of the 1173 cells in our basic data table there are 884 zero values! Clearly, such a data set is ill-conditioned for most types of statistical analysis.

The consequences regarding classical statistical analysis based on the system of moments and product moments not withstanding we have to take zero inundated data sets as normal in Vegetation Ecology. We could try to change the data to fit some available statistical technique but it would be foolhardy. The technique has to be selected to fit the data. Not the other way around.

As we already mentioned, zero values are not a hindrance in quantum analysis. The reason is as we pointed out earlier that quantum analysis uses the marginal totals (T) and the number of none zero values (n) which contribute materially to T.

ENERGY STRUCTURE

The species level

The energy states of the 23 sections in the proxy succession sere are our tangibles for the energy structure of the primary succession sere in the catena site. The T and n values required in the computation of nH and H are included in Appendix 1 or derived from the given values as needed. Table 1 contains the first set of results. In this the basic processes' energy footprints are isolated on the species level of the phylogenetic tree (see Part I of Appendix 1 for taxonomic code).

Table 1. Potential energy footprints of the basic processes: environmental sorting of functional types, phylogeny, and emergent effects on the species level in succession. All H are in nats.

Potential energy	T	n	nH	H	P
Current environmental sorting	2018	23	126.041	5.480	0.004
Phylogeny	2018	51	239.219	4.691	0.009
Emergent effects	2018	135*	505.164**	3.737	0.024
Proxy succession sere	2018	289	870.424	3.012	0.049

*Value determined by interpolation based on T and nH
** Value determined by taking the difference

Considering that the P values are significant, also in the case of the emergent effects, the existence of specific energy footprints is shown and the seamless fusion proposition (Proposition A) is accepted.

19 | Quantum analysis of succession

Probably it already crossed the reader's mind that the Devastation Trail proxy succession sere intersect the segments of an environmentally heterogeneous catena. This allowed the Kernerian process of facilitation (Kerner 1863) to show up in the analysis as a tangible in the disguise of environmental sorting of the functional types into communities.

The footprint of phylogeny on the species level has much to do with functional type richness, for which the long-term evolutionary process is responsible. The phylogenetic process intercepts the current succession process at the species level. But there are other historic effects which structure the species-level energy footprint hierarchically into components. We explore this topic in other sections.

The energy footprint under the "emergent" label cannot include the interaction of phylogeny and environmental mediation, since the two processes only meet in the present. The likely source is the category of chance events, which the Gaussian Normal spectrum accommodates, totally ubiquitous and unpredictable.

We remind the reader that the nH values of Table 1 can be interpreted locally. When comparisons are intended, the per resonator value $H = \dfrac{nH}{n}$ have to be used. Uniqueness, *i.e.* statistical significance, *is* declared when there is strong evidence that the observed H value is NOT the consequence of blind chance. The threshold criterion is H =3 nats or P=0.05. In other words, we consider H unique if it is not less than 3 nats, or equivalently, if the P value is not greater than 0.05. By this convention, we see indications of strong determinism[2] in the evolution of the potential energy structure in succession under the dynamic effects of the basic processes.

We may ask is one of the basic processes dominant? This is the same as asking if the difference of the H values is statistically significant. Considering the difference dH = 0.789, we find that it falls far below the 3 threshold for significance. We conclude that in H terms and at the species level in the phylogenetic hierarchy the potential energy footprint of environment mediated sorting and phylogeny have statistically indistinguishable energy footprints. The conclusion based on nH is quite different.

Energy hierarchy

The forgoing analysis had the species level of the phylogenetic hierarchy in its hair cross. The following analytical steps tell us what to expect if the taxonomic baseline is moved up from the species level to the genus, family, or order. The results are summarised in Table 2.

Table 2. Hierarchic energy structure of the proxy succession sere. Note: n is the number of nodes on the different levels in the phylogenetic hierarchy.

Level	T	n	nH	dnH	dn	dH	P	H	P
3	2018	21	116.981					5.571	0.004
2	2018	23	126.041	9.060	2.000	4.530	0.011	5.480	0.004
1	2018	40	197.233	71.192	17.000	4.188	0.015	4.931	0.007
0	2018	51	239.219	41.986	11.000	3.817	0.022	4.691	0.009

---

[2] The physical complement of randomness in chaos; the physical property of directedness about which chance events occur.

21 | Quantum analysis of succession

Inspection of the numbers in Table 2 reveals interesting facts and invites comments:

1. The nH energy state increases downward from a meagre 116 nats on Level 3 (Order) to the much more substantial 239 nats on Level 0 (Species). The slope of increase is about 88°. The increase is closely linked to the proliferation of taxa (n). At each step a dnH quantity is added onto nH. This dnH is specific to the level. For example dnH = 126 − 117 = 9 nats is the potential energy specific to Level 2 (Family).

2. The values we should be comparing are in the dH column. For example, $dH = \dfrac{9}{23-21} = 4.5$ nats is the per taxon potential energy increment owing to splitting the Order level taxa into Family level taxa. When we consider the values in the P column for H and dH we find that each evolutionary step contribute significantly to the energy structure of the local succession sere, albeit a slightly reduced rate.

3. We can contemplate the consequences if plant identification bottomed out on Level 3 (Order). For one thing, we would lose much precision in the estimation of phylogeny's energy footprint. The hierarchical structure of nH and H would remain unrevealed.

4. What do the potential energy increments manifest to us? They are manifestations of "work" done by the phylogenetic process when it split higher level taxa into more numerous lower level taxa. We do not see the complete logic of the process in numbers, only that limited portion which applies to the local

flora in the study site. Yet we can see enough by way of the probabilities to conclude that phylogeny was not ruled by chance; on all levels the dH and H values are statistically significant.

A further point is that the energy footprint of phylogeny unfolds at specific velocities through the hierarchical levels. These are traced by the nH and H values (Figure 3). We can say that the numerical trend down through the levels is deceleration in H terms on a slope 18° steep. The relative nominal steepness of the energy slope notwithstanding, the maximum difference of the H values through the horizontal scale do not reach 3. This negates statistically significant sloping on one taxon basis in the phylogenetic process. A completely different conclusion emerges in nH energy terms. But then the effect of n on comparability applies.

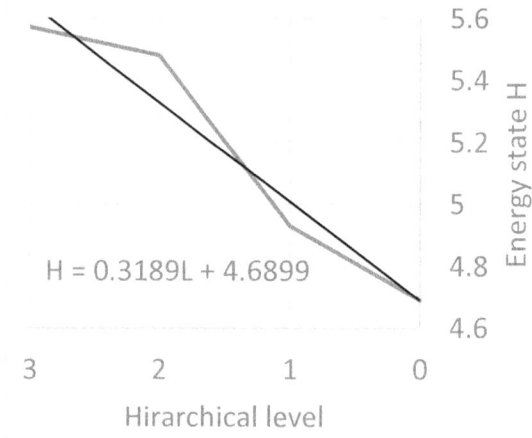

Figure 3. The H trend on declining hierarchical level.

23 | Quantum analysis of succession

Energy trends

The vegetation's energy state in each section of the catena is a predictor of what we may expect in the vegetation community at a comparable position of the chronosequence in the primary succession. In this sense, the energy structure's dynamics through succession is traceable by the set of sectional nH and H values in the catena. There are 23 of each, which we present in Table 3. Selected columns have graphs in Figure 4. Linear correlation values are placed in Table 4.

Table 3. Evolution of the energy structure of the community per section in the proxy succession sere. Legend: n – number of species, Pos - position in sere, Cov % - ground cover of all specie, MF % - ground cover of Myrica faya, MP % - ground cover of Metrosideros polymorpha.

Pos	T	n	nH	H	P	Cov %	MF %	MP %
1	11	3	7.274	2.425	0.089	10	0	0
2	11	3	7.274	2.425	0.089	10	0	0
3	12	3	7.506	2.502	0.082	10	0	0
4	12	4	8.997	2.249	0.105	10	0	0
5	21	9	18.326	2.036	0.131	10	0	0
6	40	11	26.591	2.417	0.089	20	0	5
7	59	13	34.001	2.615	0.073	20	0	10
8	71	16	41.522	2.595	0.075	30	0	10
9	93	20	52.748	2.637	0.072	40	0	15
10	104	19	52.938	2.786	0.062	50	0	20
11	126	17	52.151	3.068	0.047	60	10	20
12	139	16	51.478	3.217	0.040	60	10	20
13	127	18	54.388	3.022	0.049	60	20	25
14	132	19	57.135	3.007	0.049	70	30	20
15	153	17	55.264	3.251	0.039	70	40	60
16	116	12	39.825	3.319	0.036	70	20	70
17	96	11	35.438	3.222	0.040	70	10	65
18	152	11	40.275	3.661	0.026	80	50	80
19	130	12	41.129	3.427	0.032	80	30	65
20	106	8	28.967	3.621	0.027	70	40	60
21	88	10	32.295	3.230	0.040	60	30	50

22	97	19	51.725	2.722	0.066	60	20	60
23	122	18	53.713	2.984	0.051	70	0	80

Table 4. Linear correlation of specified entities based on Table 3. Legend: Pos – position in sere, Cov % - ground cover of all specie, MF % - ground-cover of Myrica faya, MP % - ground cover of Metrosideros polymorpha.

	Pos	T	n	nH	H	Cov %	MF %	MP %
Pos	1	0.769	0.484	0.623	0.784	0.898	0.673	0.914
T	0.769	1	0.728	0.879	0.841	0.939	0.153	0.298
n	0.484	0.728	1	0.965	0.293	0.567	0.153	0.298
nH	0.623	0.879	0.965	1	0.521	0.746	0.351	0.470
H	0.784	0.841	0.293	0.521	1	0.913	0.816	0.811
Cov &	0.898	0.939	0.567	0.746	0.913	1	0.750	0.861
MF %	0.673	0.153	0.153	0.351	0.816	0.750	1	0.687
MP %	0.914	0.298	0.298	0.470	0.811	0.861	0.687	1

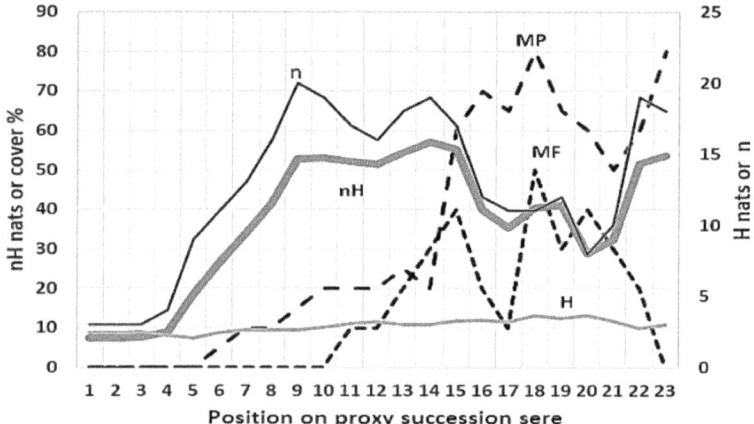

Figure 4. The nH and H trends in the proxy succession sere over 23 catena segments. Legend: n - number of species, MP % - ground cover of Metrosideros polymorpha, MF % - ground cover of Myrica faya, nH –potential energy footprint of the plant community, H - the per species share of nH.

The results suggest well-defined nH and H specific energy trends:

1. The nH graph sensitively follows species richness (n) as shown by the regression equation nH = 2.9486n - 0.051 and by the coefficient of determination R^2 = 0.9315 (Figure 5).

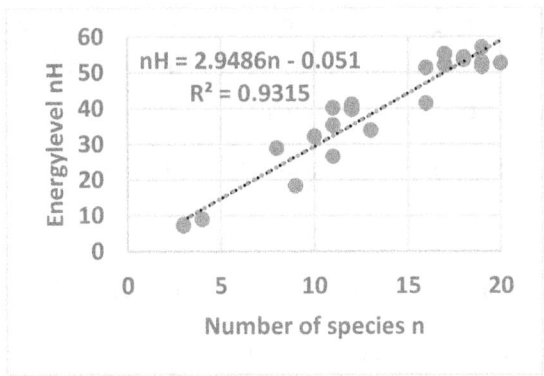

Figure 5. Regression of nH on n. See data in Table 3.

H in turn has sensitive dependence on the transient dominance of species (Figure 6). The regression equation is H = 0.0156 Cover% + 2.1483 and R^2 = 0.8334.

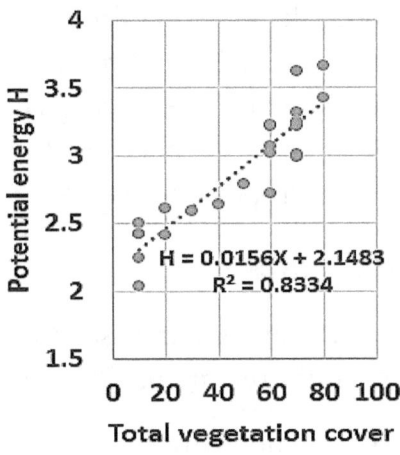

Figure 6. Regression of H on vegetation cover %. See data in Table 3.

So the key pattern variable for nH is n, whereas Cover % is the key pattern variable for H.

2. H is free of the differential effects of richness, co-varies sensitively with total ground cover, which increases with position in the proxy succession sere. The corresponding regression equations show these properties:

$$H = 0.0514 Pos + 2.2719, R^2 = 0.615$$

and

$$\text{Cover \%} = 3.4387 Pos + 6.1265, R^2 = 0.806.$$

H rises over the 23 catena sections at an angle of roughly 3°. In dH terms this corresponds to 1.2 nats rise which is less than the critical 3 nats for statistical significance. If we assume 20 species in the community in section 23 then the rise of nH is dnH = 24 nats. This is an estimate of the energy spent in the course of the succession process for the vegetation to reach the forest state.

What should we conclude regarding Proposition B? The vegetation's energy state measured in nH terms does in fact follow the proposition with ascent on a 24/23 slope in radians which is more than 45°. Obvious from Table 3 and Figure 4 that in H terms we have to reject Proposition B.

Homogeneity tests

One more phenomenon invites our closer scrutiny. It is the homogeneity of the energy structure in space and time.

We mentioned in connection with Figure 4 that the nH graph has sharp inflexions in different sections of the proxy succession

27 | Quantum analysis of succession

sere, yet the H graph is flat. One of the question that comes up when we observe the graphs is concerned with the homogeneity of the sectional energy footprints across different numbers of sections in the proxy succession sere. We can test homogeneity based on the contents of Table 5.

Table 5. Homogeneity tests of the sectional energy footprints of the proxy succession sere. Segment limits refer to section numbers from 1 to 23.

Part I: Homogeneity of segments in the 23-unit proxy sere, species level. Legend: n – number of species within the segment, nH – total energy footprint of the segment, h – the per species energy footprint of the segment.

Segment	T	n	nH	dnH	H	P	max dH	P
1 to 23	2018	51	239.2190		4.6906	0.0092		
1 to 12	958	35	151.4642		4.3275	0.0132		
12 to 23	1192	42	183.2516	31.7874	4.3631	0.0127		
1 to 9	330	25	90.4293		3.6172	0.0269		
9 to 16	990	36	155.9574	65.5281	4.3322	0.0131		
16 to 20	600	23	98.4483	57.5091	4.2804	0.0138		
20 to 23	413	28	104.2832	5.8349	3.7244	0.0241	0.715	0.5446

Part II: Homogeneity of energy footprints across the 4-level phylogenetic hierarchy within segments of the proxy succession series. Legend: n – number of levels, nH – total energy footprint of phylogeny in the segment, H – the per level energy footprint of phylogeny within the segment.

Segment	T	n	nH	dnH	H	P	max dH	P
1 to 23	2018	4	28.8982		7.2246	0.0007		
1 to 12	958	4	25.9226	2.9756	6.4806	0.0015		0.4752
12 to 23	1192	4	26.7951	0.8725	6.6988	0.0012		
1 to 9	330	4	21.6753	5.1198	5.4188	0.0044		
9 to 16	990	4	26.0537	4.3784	6.5134	0.0015		
16 to 20	600	4	24.0558	1.9979	6.0140	0.0024		
20 to 23	413	4	22.5679	1.4879	5.6420	0.0035	0.8714	0.4183

What can we conclude from Part I in Table 5? -

1. The per species potential energy footprint H of the 23 unit proxy sere is 4.6906 nats. The H values in segments 1 to 12 and 12 to 23 differ by 0.0356, certainly much less the 3 nats. We can declare the two segments homogeneous.

2. Taking the last 4 segments, the maximum difference is dH = 0.715 nats. Again this value is too small to be regarded significant. We conclude from this that the 4 segments are homogeneous.

3. If we consider values of in the dnH column of Part I of Table 5, we see that all values exceed 3. We conclude heterogeneity.

Continuing with Part II of Table 5, we consider the segments homogeneous regarding the H footprint and selectively heterogeneous regarding the nH footprints.

We now perform homogeneity tests within the hierarchical levels bounded by the specified sections in Table 6. We find no two H value differing by 3 nats. But all dnH are greater than 3. We conclude universal homogeneity based on H and universal heterogeneity on nH. What we make of this? The homogeneity of energy footprints for the phylogenetic level in H terms is retained for the full length of the segments. The heterogeneity of the phylogenetic levels in nH terms is also retained.

Table 6. Homogeneity tests within hierarchical levels bonded by catena sections. Legend: n - number of nodes within the segments, nH – total energy footprint of the level within the segment, H – the per node (taxon) energy footprint within the segment. .

Section	Level	T	n	nH	dnH	H	P
1 to 23	3	2018	21	116.981		5.571	0.004

29 | Quantum analysis of succession

	2	2018	23	126.041	9.06	5.48	0.004
	1	2018	40	197.233	71.192	4.931	0.007
	0	2018	51	239.219	41.986	4.691	0.009
1 to 12	3	958	15	77.469		5.165	0.006
	2	958	18	89.709	12.24	4.984	0.007
	1	958	24	112.781	23.072	4.699	0.009
	0	958	35	151.464	38.683	4.328	0.013
12 to 23	3	1192	21	106.000		5.048	0.006
	2	1192	22	110.033	4.033	5.002	0.007
	1	1192	37	166.050	56.017	4.488	0.011
	0	1192	42	183.252	17.202	4.363	0.013
1 to 9	3	330	12	51.986		4.332	0.013
	2	330	13	55.297	3.311	4.254	0.014
	1	330	20	76.661	21.364	3.833	0.022
	0	330	25	90.429	13.768	3.617	0.027
9 to 16	3	990	15	77.958		5.197	0.006
	2	990	17	86.241	8.283	5.073	0.006
	1	990	27	124.615	38.374	4.615	0.01
	0	990	36	155.957	31.342	4.332	0.013
16 to 20	3	600	13	62.956		4.843	0.008
	2	600	15	70.519	7.563	4.701	0.009
	1	600	21	91.764	21.245	4.37	0.013
	0	600	23	98.448	6.684	4.28	0.014
20 to 23	3	413	20	81.031		4.052	0.017
	2	413	21	84.083	3.052	4.004	0.018
	1	413	26	98.701	14.618	3.796	0.022
	0	413	28	104.283	5.582	3.724	0.024

Our final conclusion is that primary succession generates heterogeneity in nH terms, but leaves homogeneity untouched in H terms. These suggest in an indirect way the rule of species richness over nH type heterogeneity.

DISCUSSION

Quantum analysis has been introduced recently in vegetation studies. Not well known and for that reason a review of the criteria put forward to be considered is appropriate:

1. Quantum analysis supports a holistic approach in energetics.

2. Quantum analysis as we use it shifts the focus from conventional calorific (trophic) flow in the ecosystem to the potential energy structure of the community processes.

3. The energy model of quantum analysis is parameterised by the usual type of survey and experimental data recorded in vegetation studies.

The energy structure of the community we are interested in when we do quantum analysis is a seamless fusion of the energy footprints of the basic processes which govern vegetation dynamics. In other words, the model is not a map of calorific flow from and to existing components, but a map of connectedness to ancestors in the far past in which the phylogenetic tree is our guide, to the current environmental mediation in the sorting of functional types into communities, and to the ubiquitous effects of chance present everywhere and at all times in the vegetation process.

We call the vegetation a well-structured universe, a complex. Our aim with quantum analysis is to lay this complex's energy structure open to reveal energy patterns and trends in time and space. Quantum analysis does the job through isolation of the basic processes' footprints and tracing their dynamics in time and space.

The energy parameter in quantum analysis is Max Planck's energy-based entropy function. In this, "entropy" is just another name for "energy", which is potential energy, measured by the nH and H quantities. These are probability based measures in the manner of $nH = -\ln P$. P is the probability that the complex being exactly what it is by pure chance. Since we can determine P for any vegetation complex, we can determine also its potential energy state.

The total succession process is certainly a many faceted object and for that reason the potential energy state of the complex at a point in the process will appear the way it actually does, depending on the assembled resonators. Therefore, whether the energy state is low or high is meaningful only in comparative terms. Irrespective of its amount, the potential energy structure is the outcome of work. We rarely have the advantage of knowing what kind of work. We only see the outcome in the patterns and trends of nH and H we determine.

So what do we really gain by knowing the potential energy state of a vegetation community? We certainly do not measure "energy" in calorific terms for which any one would see immediate utility, and for that reason a need for knowing it. What is then the utility in knowing Max Planck potential energy of a resonator complex? Physics had no problem to find utility for it. What about ecology?-

33 | Quantum analysis of succession

1. Think of $nH = \ln C$ and call it a measure of the state of complexity in multispecies communities.

2. Think of nH or H as an indicator of uniqueness.

3. Construct a multiscale, hierarchical statistical approach around nH and H for data analysis and use it to test conjectures about complexity or uniqueness.

4. Consider nH or H as a holistic, high level diversity measure, a supplement to Rényi's generalised entropy.

5. Above all, think of H as a gateway to a holistic alternative to ecological energetics with focus on potential energy.

The diligent student will no doubt try to expand on the items in the list and find ways to amend the list with new items. He or she may even try to look into the utilities of nH and H in physics. Only one thing they will not do: miss the point that the individual resonators' p_i values, for which Rényi's generalized entropy is formulated, are not considered as such in quantum analysis.

The nH and H are quantities readily converted to probabilities by the negative exponential function. H being the energy state of a single resonator and nH being the energy state of the complex, they are not expected to lead to the same conclusion in probabilistic tests. We mentioned the dilemma on this count and put the question already:

"Can the nH vs. H dilemma be resolved? "

The nH energy quantity applies to the complex. Therefore on the basis of nH we can tell what is the size of the energy footprint of the complex, and if it suggests chance effects at work or determinism. H being n neutral, makes the conclusions about energy footprints not only universally comparable, but also conservative with a bias against accepting falls propositions.

The above conclusion leads us to what may be a revelation to the reader about quantum analysis:

The species are taken as objects, nothing more and nothing less than any physical resonator which possess potential energy in proportions of their performance, shape, and mass by way of $P = \dfrac{1}{C}$. The value of H and nH being the consequence of the attribute actually measured, and not unmeasured, leads us to the re-revelation that what we see is what the scale allows us to see. Nothing in ecological studies I can think of is absolute. Scale variables are everywhere and they define ecological perceptions. Some of the scale variables in study include catena section size and length, cover %, level to which the taxa are identified, highest and lowest taxonomic bound, dimension of the energy unit, and still others. Therefore our first step in the presentation of results must always be the clarification of scales within which the conclusions have their validity.

Whatever trait we measure on the species, cover % or something else, the magnitude of the trait to be achieved requires energy to be put to work. The energy is the potential energy we gauge by nH or H in nats. This is just as valid a unit if chosen with forethought of what is to be achieved as anything else if

35 | Quantum analysis of succession

called for by good reason. This has direct analogy with our earlier example in which we elected to measure potential energy in hkgs^{-2} units.

References

Brillouin, L. 1962. Science and Information Theory. 2nd ed. Academic Press, New York.

Hawking, S. (ed). 2011. The dreams that stuff is made of. Running Press, London.

Kerner von Marilaun, A. 1863. Das Pflanzenleben der Donauländer. Innsbruck, Wagner. – English rewrite by Conard, H.S. 1951. The Background of Plant Ecology. The Iowa State University Press, Ames (1977, Arno Press, New York.).

Kitayama, K., D. Mueller-Dombois and P.M. Vitousek. 1995. Primary succession of Hawaiian montane rain forest on a chronosequence of eight lava flows. Journal of Vegetation Science 6: 211-222.

Mueller-Dombois, D. and F.R. Fosberg. 1998. Vegetation of the Tropical Pacific Island. Springer.

Mueller-Dombois, D., Jacobi, J.D, Boehmer, H.J., J.P Price. 2013. 'Ohi'a Lehua Rainforest. Born among Hawaiian volcanoes, evolved in isolation. Josef Rock herbarium, Honolulu. https://www.createspace.com/3942792

Odum, H.T. 1971. Environment, Power, and Society. Wiley-Interscience, New York.

Orlóci, L. 2006. Diversity partitions in 3-way sorting: functions, Venn diagram mappings, typical additive series, and examples. Community Ecology 7: 253-259.

Orlóci, L. 2012. Self-organisation and Mediated Transience in Plant Communities. SCADA Publishing, Canada. Enlarged Online Edition: https://createspace.com/3585127

Orlóci, L. 2012. Statistical Ecology. The quantitative exploration of nature to reveal the unexpected. SCADA Publishing, Canada. Online Edition: https://createspace.com/3476529

Orlóci. L. 2013. Quantum Ecology. Energy structure and its analysis. SCADA Publishing, Canada. Online Edition: https://cratespace.com/4406077

Planck, Max. 1901. On the law of distribution of energy in the normal spectrum. Annalen der Physik Vol. 4, p. 553 et seq.

Rényi, A. 1961. On measures of entropy and information. In: J. Neyman (ed.), Proceedings of the 4th Berkeley Symposium on Mathematical Statistics and Probability, pp. 547-561. University of California Press, Berkeley.

Shannon, C. E. 1948. A mathematical theory of communication. Bell System Tech. J. 27: 379-423.

Smathers, A.G. and D. Mueller-Dombois. 2007. Hawai'i, the fires of life. Honolulu, Mutual Publishing.

Wildi, O. and M. Schüts. 2000. Reconstruction of a 405 yr. recovery process from pasture to forest. Community Ecology 1: 25-32.

Index

Andropogon virginicum, 11
Appendices, 41
baseline, 20
basic processes, 2, 6, 7, 18, 20, 31, 32
Bibliographic notes, 55
Boehner, 8, 36
Brillouin, 16, 36
calorific (trophic) flow, 31
catena, 8, 9, 11, 13, 14, 15, 18, 19, 23, 24, 26, 34, 41
catena site, 8, 18
chance, 2, 6, 7, 19, 20, 22, 31, 32, 34
characteristics sections, 8
chronosere, 1, 2, 3, 6, 8, 9
Cibotium glaucum, 11
coefficient of determination, 25
complex, 13, 14, 31, 32, 33, 34
composite proxy succession sere, 9
Conard, 36
context dependence, 14
correlation, 23, 24

determinism, 20, 34
Devastation Trail, 8, 9, 19, 41
diversity analysis, 15
Dubautia scabra, 11
dynamics, 9, 23, 31, 32
ecology, 4, 9, 32, 55
emergent effects, 18
energy, 1, 3, 6, 7, 8, 12, 13, 14, 15, 17, 18, 19, 20, 21, 22, 23, 24, 26, 27, 28, 31, 32, 33, 34, 37, 55
energy footprint, 7, 19, 20, 21, 22, 24, 27, 28, 34
energy footprints, 7, 8, 18, 20, 27, 28, 31, 34
energy state, 7, 11, 12, 13, 14, 15, 21, 23, 26, 32, 33
energy structural, 6
energy structure, 1, 2, 3, 6, 7, 15, 18, 20, 21, 23, 26, 31, 32, 55
energy units counted, 13
entropy function, 6, 13, 16, 32
environmental mediation, 2, 6, 7, 19

39 | Quantum analysis of succession

family, 11, 20, 55
for time substitution, 8
Fosberg, 9, 36
functional type, 19
functional types, 7, 18, 19, 31
Gaussian Normal spectrum, 19
genus, 20
ground cover, 8, 23, 24, 26
H, 13, 14, 15, 16, 17, 18, 19, 20, 21, 22, 23, 24, 25, 26, 27, 28, 30, 32, 33, 34, 36
Hawai'i Volcanos National Park, 1, 2, 3, 6, 8, 9
Hawaiian Islands, 8
Hawking, 15
Hedychium gardnerianum, 11
hierarchical structure, 21
holistic approach, 31
homogeneity, 26, 27, 28, 30
Homogeneity tests, 26, 27, 28
identification, 21
Information, 36
Jacobi, 8, 36
Kerner, 36
Kitamaya, 9
lava flows, 9, 36
Márta Mihály, 8, 55
Mauna Loa, 9
Max Planck, 6, 11, 12, 13, 14, 15, 32
Metrosideros polymorpha, 11, 23, 24

Metrosideros Rainforest Biome, 8
model, 31
Mueller-Dombois, 8, 36, 37, 41
multiscale, 6, 11, 33
multiscaling, 4
Myrica faya, 11, 23, 24
natural history, 9
natural units, 13
Nephrolepis exaltata, 11
nH, 13, 14, 17, 18, 19, 20, 21, 22, 23, 24, 25, 26, 27, 28, 30, 32, 33, 34
nH vs. H dilemma, 33
nodes, 20, 28, 41
Odum, 36
Orlóci, 3, 4, 11, 12, 13, 14, 15, 16, 37, 55
phylogenetic hierarchy, 20, 27
phylogenetic tree, 18, 31
phylogeny, 2, 6, 7, 18, 19, 20, 21, 22, 27
phytosociological records, 8
potential energy, 2, 6, 7, 11, 12, 15, 20, 21, 24, 28, 31, 32, 33, 34
Price, 8, 36
primary succession, 2, 6, 7, 9, 18, 23, 30
principles, 11, 12, 13
propositions, 7, 11, 34
proxy succession sere, 6, 8, 18, 19, 20, 23, 24, 26, 27

quanta, 14
quantum, 2, 6, 8, 11, 12, 13, 15, 16, 17, 31, 32, 33, 34
quantum analysis, 2, 6, 8, 11, 12, 15, 16, 17, 31, 32, 33, 34
Quantum analysis, 15, 31, 32
quantum theory, 11, 12
Rényi, 16, 33, 37
resonators, 13, 14, 32, 33
richness, 7, 14, 19, 25, 26, 30
sample plots, 9
Scale variables, 34
seamless fusion, 2, 6, 7, 31
Shannon, 16
slope, 8, 21, 22, 26
Smathers, 8, 37
space for time substitution, 9
spatial chronosere, 8
species, 6, 7, 8, 11, 14, 15, 18, 19, 20, 23, 24, 25, 26, 27, 28, 30, 34, 41
Stirling's approximation, 13
Styphelia douglasii, 11
succession, 6, 7, 8, 9, 15, 18, 19, 20, 21, 23, 24, 26, 27, 32, 36
succession process, 7, 19, 26, 32
synthetic chronoseres, 8
T, 11, 12, 14, 16, 17, 18, 20, 23, 24, 27, 28, 36, 41, 44
taxonomic code, 18
trajectory, 55
ubiquitous effects, 31
uniqueness, 15, 33
universal homogeneity, 28
Vaccinium ciliatum, 11
vegetation, 1, 2, 3, 4, 6, 7, 8, 9, 11, 13, 14, 15, 17, 23, 25, 26, 31, 32, 55
vegetation process, 4, 31
Veronica, 11
Vitousek, 9, 36
work, 9, 21, 32, 34
zero inundated data, 11, 17

Appendices

Table 1. The Devastation Trail catena's data set. The site is in the Hawaii Volcanoes National Park. The table is structured in three parts. Species names and systematic code are given in Part I. Species totals are given for different segments of the catena in the last seven columns of the Part II. Species names follow usage in the local scientific literature (Mueller-Dombois et al. 2013). Segment position, segment totals and species numbers are listed in Part III. Symbols: T total cover, n number of nodes.

Part I

#	Species	Genus	Code	Family	Code	Order	Code
22	Metrosideros polymorpha var. poly.	Metrosideros	21	Myrtaceaea	10	Myrtales	9
16	Myrica faya	Myrica	22	Myricaceae	10	Fagales	6
14	Vaccinium reticulatum	Vaccinium	39	Ericaceae	5	Ericales	4
24	Andropogon virginicus	Andropogon	2	Poaceae	14	Poales	10
49	Nephrolepis exaltata	Nephrolepis	23	Lomariopsidaceae	8	Polypodales	20
20	Veronica (piros foltok)	Veronica	40	Plantaginaceae	13	Lamiales	8
6	Dubautia ciliolata	Dubautia	14	Asteraceae	1	Asterales	2
3	Bidens alba var. radiata	Bidens	5	Asteraceae	1	Asterales	2
19	Buddleja asiatica	Buddleja	6	Scrophulariaceae	22	Lamiales	8
51	Sadleria cyantheoides	Sadleria	33	Blechnaceae	2	Blechnales	21
17	Coprosma ernodeoides	Coprosma	10	Rubiaceae	20	Gentianales	7
42	Dodonea viscosa	Dodonaea	13	Sapindaceae	21	Sapindales	13
9	Rumex giganteus	Rumex	32	Polygonaceae	15	Caryophyllales	3
10	Styphelia douglasii	Styphelia	37	Ericaceae	5	Ericales	4
44	Psilotum nudum	Psilotum	29	Psilotaceae	16	Psilotales	15
21	Metrosideros polimorpha var. glab.	Metrosideros	21	Myrtaceaea	10	Myrtales	9
37	Rubus penetrans	Rubus	31	Rosaceaea	19	Rosales	12
34	Anemone hupehensis var. japonica	Anemone	3	Ranunculaceae	18	Ranunculales	11
47	Cibotium glaucum	Cibotium	9	Cibotiaceae	3	Cyatheales	18
13	Vaccinium peleanum	Vaccinium	39	Ericaceae	5	Ericales	4
18	Coprosma ochracea var. rockiana	Coprosma	10	Rubiaceae	20	Gentianales	7
41	Epilobium billardierianum subsp. cin.	Epilobium	15	Onograceae	11	Rosales	12
43	Lycopodium cernuum	Lycopodium	19	Lycopodiaceae	9	Lycopodiales	14

#	Species	Genus		Family		Order	
4	Bidens pilosa	Bidens	5	Asteraceae	1	Asterales	2
5	Dubautia scabra	Dubautia	14	Asteraceae	1	Asterales	2
7	Hypochoeris radicata	Hypochaeris	17	Asteraceae	1	Asterales	2
23	Andropogon glomeratus	Andropgon	2	Poaceae	14	Poales	10
1	Arundina bambusaefolia	Arundina	4	Orchidaceae	12	Asparagales	1
2	Spathoglottis plicata	Spathoglottis	36	Orchidaceae	12	Asparagales	1
8	Sonchus oleraceus	Sonchus	35	Asteraceae	1	Asterales	2
12	Vaccinium calycinum	Vaccinium	39	Ericaceae	5	Ericales	4
26	Carex wahuensis	Carex	8	Cyperaceae	4	Poales	10
28	Kyllinga brevifolia (Cyperus)	Killinga	18	Cyperaceae	4	Poales	10
29	Matchaerina angustifolia	Matchaerina	20	Cyperaceae	4	Poales	10
30	Paspalum conjugatum	Paspalum	25	Poaceae	14	Poales	10
32	Pycreus polystachios (Cyperus)	Pycreus	30	Cyperaceae	4	Poales	10
38	Rubus rosifolius	Rubus	31	Rosaceaea	19	Rosales	12
39	Touchardia latifolia	Touchardia	38	Urticaceae	23	Rosales	12
40	Fragaria vesca var. alba	Fragaria	16	Rosaceae	19	Rosales	12
48	Dicranopteris linearis	Dicranopteris	12	Gleicheniacea	7	Cyatheales	19
11	Styphelia tameiameiae	Styphelia	37	Ericaceae	5	Ericales	4
15	Acacia koa	Acacia	1	Fabaceae	6	Fabales	5
25	Bulbostylis capillaris	Bulbostylis	7	Cyperaceae	4	Poales	10
27	Cyperus rotundifolius	Cyperus	11	Cyperaceae	4	Poales	10
31	Paspalum dilatatum	Paspalum	25	Poaceae	14	Poales	10
33	Setaria geniculate	Setaria	34	Poaceae	14	Poales	10
35	Osteomeles anthyllidifolia	Osteomeles	24	Rosaceae	19	Rosales	12
36	Pipturus albidus	Pipturus	26	Urticaceae	23	Rosales	12
45	Pityrogramma calomelanus	Pityrogramma	27	Pteridaceae	17	Pteridales	16
46	Polypodium pellucidum	Polypodium	28	Polypodiaceae	15	Polypodales	17
50	Nephrolepis hirsutula	Nephrolepis	23	Lomariopsidaceae	8	Polypodales	20
		T					
		n	40		23		21

Part II

#	Species	1 to 23	1 to 12	12 to 23	1 to 9	9 to 16	16 to 20	20 to 23
22	Metrosideros polymorpha var.poly.	735	145	610	40	250	340	250
16	Myrica faya	310	70	270	0	130	150	90
14	Vaccinium reticulatum	121	100	26	55	65	13	4
24	Andropogon virginicus	116	98	28	23	100	6	2
49	Nephrolepis exaltata	104	78	41	18	76	9	8
20	Veronica (piros foltok)	101	90	21	25	85	5	1

Quantum analysis of succession

6	Dubautia ciliolata	79	58	26	33	36	13	4
3	Bidens alba var. radiata	58	55	8	10	52	2	0
19	Buddleja asiatica	53	52	11	12	45	0	1
51	Sadleria cyantheoides	42	38	9	12	29	1	2
17	Coprosma ernodeoides	40	20	20	15	15	10	5
42	Dodonea viscosa	33	0	33	0	1	21	12
9	Rumex giganteus	32	32	0	30	7	0	0
10	Styphelia douglasii	27	22	5	17	6	2	4
44	Psilotum nudum	22	19	4	2	19	0	2
21	Metrosideros sp	20	5	20	0	15	10	0
37	Rubus penetrans	20	18	7	5	20	1	0
34	Anemone hupehensis var. japonica	18	16	3	15	1	0	2
47	Cibotium glaucum	8	1	7	0	1	1	6
13	Vaccinium peleanum	7	2	5	2	1	0	5
18	Coprosma ochracea var. rockiana	7	7	1	5	7	0	0
41	Epilobium billardierianum subsp. Cin.	7	7	1	2	6	0	0
43	Lycopodium cernuum	6	0	6	0	0	5	1
4	Bidens pilosa	4	3	1	1	4	0	0
5	Dubautia scabra	4	4	0	1	3	0	0
7	Hypochoeris radicata	4	1	4	0	2	2	0
23	Andropogon glomeratus	3	3	0	2	1	0	0
1	Arundina bambusaefolia	2	0	2	0	1	1	0
2	Spathoglottis plicata	2	0	2	0	0	0	2
8	Sonchus oleraceus	2	1	1	0	1	0	1
12	Vaccinium calycinum	2	2	0	2	0	0	0
26	Carex wahuensis	2	1	1	0	2	0	0
28	Kyllinga brevifolia (Cyperus)	2	0	2	0	0	2	1
29	Matchaerina angustifolia	2	0	2	0	0	0	2
30	Paspalum conjugatum	2	2	1	0	2	0	0
32	Pycreus polystachios (Cyperus)	2	0	2	0	0	2	1
38	Rubus rosifolius	2	2	1	0	2	0	0
39	Touchardia latifolia	2	2	0	1	1	0	0

#	Species							
40	Fragaria vesca var. alba	2	0	2	0	0	2	0
48	Dicranopteris linearis	2	0	2	0	0	0	2
11	Styphelia tameiameiae	1	1	0	1	1	0	0
15	Acacia koa	1	0	1	0	0	0	1
25	Bulbostylis capillaris	1	0	1	0	0	1	0
27	Cyperus rotundifolius	1	0	1	0	0	1	0
31	Paspalum dilatatum	1	1	0	0	1	0	0
33	Setaria geniculate	1	1	0	0	1	0	0
35	Osteomeles anthyllidifolia	1	0	1	0	0	0	1
36	Pipturus albidus	1	0	1	0	0	0	1
45	Pityrogramma calomelanus	1	0	1	0	0	0	1
46	Polypodium pellucidum	1	0	1	0	0	0	1
50	Nephrolepis hirsutula	1	1	0	1	1	0	0

	T						
	T	2018	958	1192	330	990	600	413
Systematic level:	Species n	51	35	42	25	36	23	28
	Genus n	40	24	37	20	27	21	26
	Family n	23	18	22	13	17	13	21
	Order n	21	15	21	14	18	17	20

Part III.

PSS	1	2	3	4	5	6	7	8	9	10	11
T	11	11	12	12	21	40	59	71	93	104	126
n	3	3	3	4	9	11	13	16	20	19	17

	12	13	14	15	16	17	18	19	20	21	22	23
	139	127	132	153	116	96	152	130	106	88	97	122
	16	18	19	17	12	11	11	12	8	10	19	18

45 | Quantum analysis of succession

Supplementary references

Quantum analysis of primary succession: The energy structure of a vegetation chronosere in Hawaii Volcanoes National Park

Authored by Laszlo Orlóci FRSC

List Price: **$30.00**

6" x 9" (15.24 x 22.86 cm)
Black & White on White paper
54 pages

ISBN-13: 978-1492788997 (CreateSpace-Assigned)
ISBN-10: 1492788996
BISAC: Science / Life Sciences / Ecology

The book revisits the classical idea that the potential energy structure of primary succession is a seamless fusion of foot-prints specific to basic processes which operate on all scales – phylogeny, environmental mediation, and chance. The idea is tested in quantum analysis of a vegetation chronosere in Hawai'i Volcanoes National Park. How is the test constructed? What are the outcomes? What do the results tell about primary succession not already known from other sources? Stated in the briefest of terms the test re-quires temporal species performance data...

ORDER FROM CREATESPACE ESTORE:
https://www.createspace.com/4452597

Quantum ecology: Energy structure and its analysis

Authored by László Orlóci FRSC

List Price: $30.00

6" x 9" (15.24 x 22.86 cm)
Black & White on White paper
72 pages

ISBN-13: 978-1492183297
ISBN-10: 1492183296
BISAC: Science / Life Sciences / Ecology

Ecology joins forces with quantum theory on the pages of "Quantum Ecology" to create a holistic approach in energy studies.

The infusion of quantum theoretical principles allows the study focus of ecological energetics to shift from the conventional calorific (trophic) flow in ecosystems to the potential energy structure of the vegetation. The books contents cover the theory and techniques in a unique account centred on the energy equation. The equation's component terms define energy footprints specific to ecology's basic processes, such as historic phylogeny, current environmental mediation of transience, and chance. What gives practical value to quantum analysis is its ability to be parameterised by the usual type of survey or experimental data.

The book is offered for classroom use in advanced courses and technical support in research projects.

ORDER FROM CREATESPACE ESTORE:
https://www.createspace.com/4406077

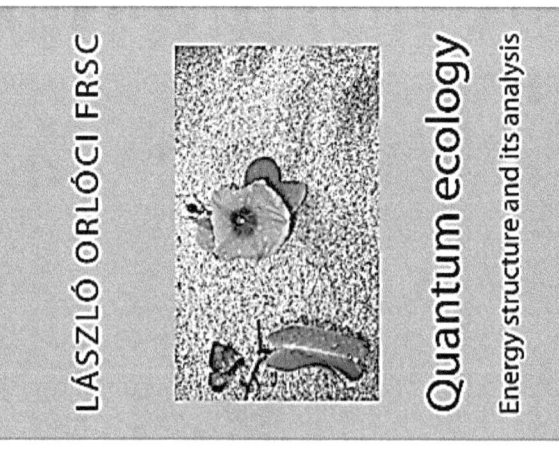

47 | Quantum analysis of succession

Statistical ecology

The quantitative exploration of Nature to reveal the unexpected

Authored by Laszlo Orlóci Ph.D.

The book's topics traverse many problem areas in univariate and multivariate data analysis, focussed on current ecological practice. The manner of presentation emphasizes reasoned methodological choices and encourages innovations consistent with the objectives, but mindful of the need to see clearly the regularity conditions in Ecology. The main text is accompanied by external appendices including a technical manual, 47 specialized application programs, and many data files taken from the exercises in the main text. For information please contact: lorloci@uwo.ca

About the author:
Orlóci is an INTECOL Distinguished Statistical Ecologist. He is external (academician) Member of the Hungarian Academy of Sciences, and regular (academician) Fellow of the Academy of Sciences of the Royal Society of Canada. He published over 100 papers in scientific journals, numerous monographs and books. His current essays on trajectory analysis, the rules of process governance, and the phylogenetic signal in vegetation transitions have considerable significance for evolutionary ecology and global change science. His present work on energy structures in metacommunities is seminal, pointing to a new direction.

List Price: $49.90

Add to Cart

Publication Date: Aug 10 2010
ISBN/EAN13: 1453760520 / 9781453760529
Page Count: 372
Binding Type: US Trade Paper
Trim Size: 6" x 9"
Language: English
Color: Black and White
Related Categories: Science / Life Sciences / Ecology

Statistical multiscaling in dynamic ecology

 0

Probing the long-term vegetation process for patterns of parameter oscillation

Authored by László Orlóci Ph.D.

The Book's conceptualisation of multiscaling theory presents the Next Step in the study of the long-term vegetation process. The context is statistical and the process generating events have proxy in the compositional transitions of the palynological spectra. Familiarity with multiscaling is not a prerequisite. The reader shall learn from the examples how multiscaling techniques helped to identify the self-similar (fractal) nature of the process, isolate low and high instability phases, locate hotspots of compositional transitions, and link these to delayed climatic effects. He or she shall also learn how to gauge process homeomorphy among sites, isolate the random and directed effects found braided into the process, and do much more within a broad yet formal probabilistic framework. The Book's contents are taken in part from a graduate course offered in the Ecology program at UFRGS in Porto Alegre, Brazil. The examples use palynological spectra from sites on the Hungarian Great Plain and in the adjacent Carpathian Mountains. Application programs are available from the author.

Publication Date:	Mar 15 2012
ISBN/EAN13:	1475071388 / 9781475071382
Page Count:	96
Binding Type:	US Trade Paper
Trim Size:	6" x 9"
Language:	English
Color:	Black and White
Related Categories:	Science / Life Sciences / Ecology

List Price: $30.00

Add to Cart

49 | Quantum analysis of succession

Self-organization and mediated transience in plant communities

What are the rules?
Authored by Dr. László Orlóci FRSC

A novel, signal theoretical solution is sketched out for the ecological problem of how to identify and quantitatively express the assembly rules of plant communities. A case study for testing the solution leads to the astonishing conclusion that the phylogenetic signal outperforms the current environmental signal in intensity close to 7 to 1. This indicates high stability and low inclination to environment mediated transience in the community.

About the author:
László Orlóci was born in Hungary in 1932. He holds degrees in forest engineering (DFE Sopron), forestry science and biology (BSF, MSc, PhD University of British Columbia), and DSc h.c. in biology (University of Trieste). Orlóci held appointments as NATO Science Fellow (University College of North Wales), professor (University of Western Ontario), and visiting professor at universities in the Americas, the Pacific, Asia, and Europe. He is an INTECOL Distinguished Statistical Ecologists, external (academician) member of the Hungarian Academy of Sciences, and regular Fellow of the Academy of Sciences of the Royal Society of Canada.

Publication Date: Nov 11 2011
ISBN/EAN13: 1461028221 / 9781461028222
Page Count: 70
Binding Type: US Trade Paper
Trim Size: 6" x 9"
Language: English
Color: Black and White
Related Categories: Science / Life Sciences / Ecology

List Price: $25.00

[Add to Cart]

On the energy structure of natural vegetation

In search for community governance rules

Authored by Laszlo Orloci FRSC

Briefly about the book....

Vegetation Science meets quantum theory in the energy-based entropy model of this book. The model is based on Max Planck's postulate that potential energy and entropy are interchangeable parameters in resonator complexes. What is a typical outcome of the model in vegetation studies? The model hands users a set of entropy estimates and probabilities based on which the strength and uniqueness of a metacommunity's energy structure can be characterised in comparative terms.

About the author:
Orlóci is an INTECOL Distinguished Statistical Ecologist. He is external (academician) Member of the Hungarian Academy of Sciences, and regular (academician) Fellow of the Academy of Sciences of the Royal Society of Canada. Orlóci published over 100 papers in scientific journals, numerous monographs, books and book chapters. His current essays on trajectory analysis, the rules of process governance, and the phylogenetic signal in vegetation transitions have considerable significance for evolutionary ecology and global change science. His present work on energy structures in metacommunities is seminal, pointing to a new direction.

List Price: $30.00

Publication Date:	Jan 30 2013
ISBN/EAN13:	1482319373 / 9781482319378
Page Count:	46
Binding Type:	US Trade Paper
Trim Size:	6" x 9"
Language:	English
Color:	Black and White
Related Categories:	Science / Life Sciences / Ecology

51 | Quantum analysis of succession

Flexible computing in statistical ecology

External appendix to accompany L. Orlóci's Statistical Ecology
Authored by Dr. László Orlóci

Problem flexible computing in statistical ecology

The Book describes more than 40 executable (.exe) computer programs and presents examples of application which correspond to the examples included in Statistical Ecology*. The programs are flexibly problem specific and conversational. They allow option-driven selective access to specific statistical tasks. Linkages are possible through standard output and input. The description includes in each case a brief introduction, a record of the start up dialogue, and detailed record input and output sets. The source code is in True Basic. The programs are compiled and linked on a 32 bit Windows XP system and tested up to Windows 7.
The executable program library, data files and a note to users are distributed free of charge to purchasers of the Technical Manual. Requests for download information should be directed to the URL address lorloci@uwo.ca. Prior to running the application programs, installation of a recent version of True Basic (see Internet for sources) on the user's system is strongly advised.
* Orlóci, L. 2010. Statistical Ecology. The quantitative exploration of nature to reveal the unexpected. Scada Publishing, Online Edition. Copies are available from the distributor
https://www.createspace.com/3476529

Publication Date:	Apr 05 2011
ISBN/EAN13:	1460972953 / 9781460972953
Page Count:	142
Binding Type:	US Trade Paper
Trim Size:	6" x 9"
Language:	English
Color:	Black and White
Related Categories:	Science / Life Sciences / Ecology

List Price: $30.00

Add to Cart

Statistical Ecology. A reasoned approach.

53 | Quantum analysis of succession

Bibliographic notes

- László Orlóci

László Orlóci was born into a military family in Hungary in 1932. Orlóci holds degrees in forest engineering (DFE Sopron), forestry science and biology (BSF, MSc, PhD University of British Columbia), and DSc *h.c.* in biology (University of Trieste). Orlóci held appointments as NATO Science Fellow (University College of North Wales), professor (University of Western Ontario), and visiting professor at universities in the Americas, the Pacific, Asia, and Europe. He is INTECOL Distinguished Statistical Ecologists, external (academician) member of the Hungarian Academy of Sciences, and regular Fellow of the Academy of Sciences of the Royal Society of Canada.

Orlóci published over 100 papers in scientific journals, numerous monographs and books. His current essays on trajectory analysis, the rules of process governance, the phylogenetic signal in vegetation transitions, and the energy structure of the vegetation have considerable significance for evolutionary ecology and global change science.

Orlóci is married to Márta Mihály, a Sopron forest engineering alumna. Their daughter Martha Barbara is a Geography graduate of Western University, granddaughter Kathryn Orlóci-Goodison is enrolled in forestry at Lake Head University in Thunder Bay, Ontario, and granddaughter Ruth Orlóci-Goodison attends high school in Cambridge, Ontario.

2013-09-28

www.ingramcontent.com/pod-product-compliance
Lightning Source LLC
Chambersburg PA
CBHW071641170526
45166CB00003B/1381